CONNECTIONS

Improving the Environment

Maggie Freeman

Series editor Sue Palmer

OXFORD
UNIVERSITY PRESS

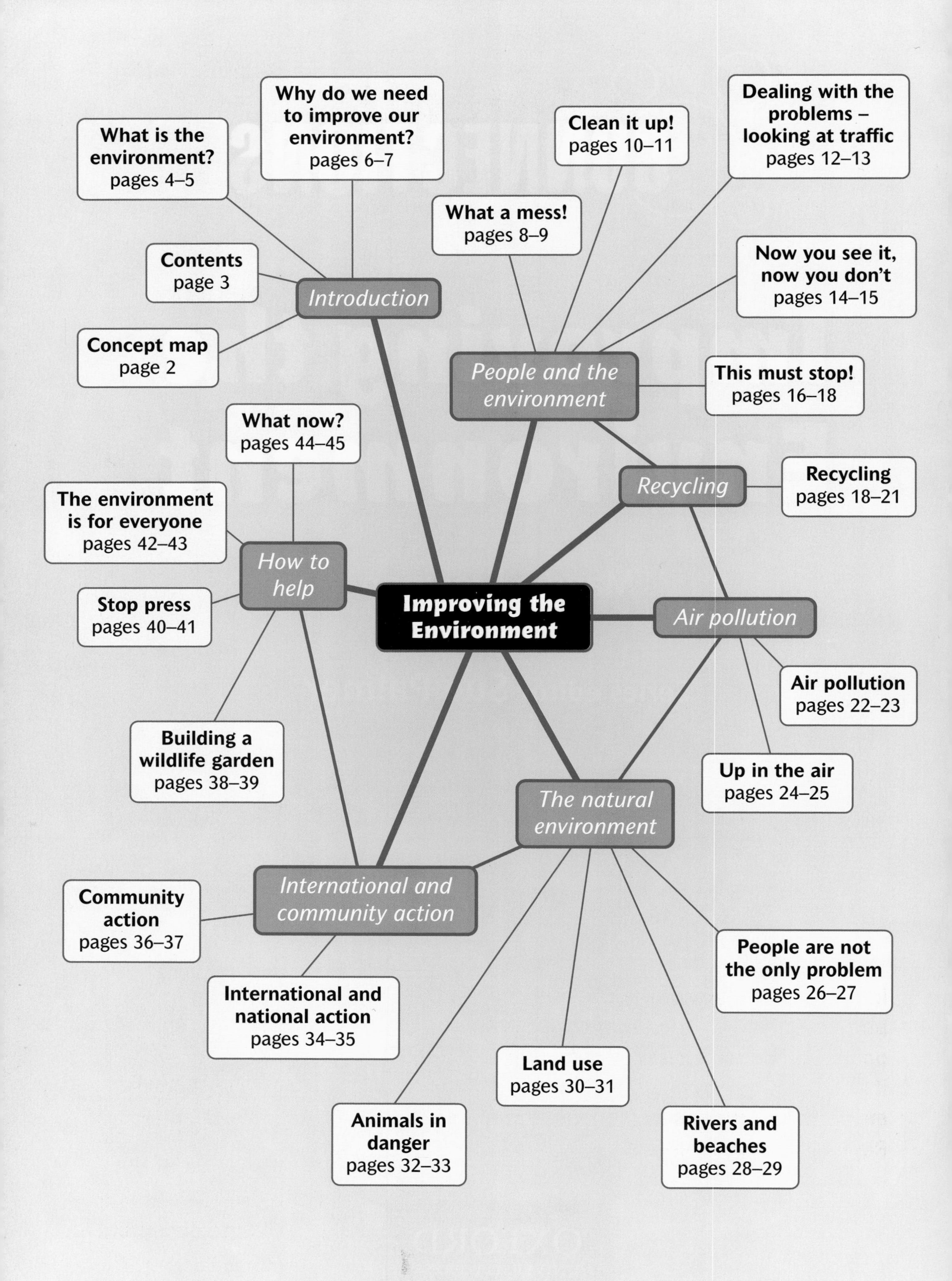

Improving the Environment
Introduction
What is the environment? pages 4–5
Why do we need to improve our environment? pages 6–7
Contents page 3
Concept map page 2
People and the environment
What a mess! pages 8–9
Clean it up! pages 10–11
Dealing with the problems – looking at traffic pages 12–13
Now you see it, now you don't pages 14–15
This must stop! pages 16–18
Recycling
Recycling pages 18–21
Air pollution
Air pollution pages 22–23
Up in the air pages 24–25
The natural environment
People are not the only problem pages 26–27
Rivers and beaches pages 28–29
Land use pages 30–31
Animals in danger pages 32–33
International and community action
International and national action pages 34–35
Community action pages 36–37
How to help
Building a wildlife garden pages 38–39
Stop press pages 40–41
The environment is for everyone pages 42–43
What now? pages 44–45

Contents

In my job, I am very aware of the environmental problems facing us today. This book will not only help you understand environmental issues, it also suggests ways you can look after the environment, such as by recycling containers. Remember – every little helps!

Mark Wells
Education Officer with SEPA (Scottish Environment Protection Agency)

What is the environment?

The environment is the world around us, and around every other form of life. When we talk about 'the environment', we mean the whole of the natural world. This includes all the small environments that belong to every single form of life on Earth. It stretches right up into the **atmosphere**, higher than a bird can fly.

Still air above a field, for a swallow hunting insects

A huge blue sky, for a soaring eagle

Air

A warm moonlit garden, for a moth

A great spacious ocean, for an enormous whale

Water

A chilly iceberg, for a polar bear cub beside her mother.

A little pond, for a hopping frog

The environment

Natural environments

Earth

A cool burrow safe from foxes, for a furry rabbit

A hot shadowy forest where he's camouflaged, for a stripy tiger

A wild wood where no woodcutters come, for old oak trees

Built environments

A busy school, for hard-working children

A cage with space to play, for an energetic hamster

A small glass tank, for bored goldfish

Why do we need to

For thousands of years humans have been affecting the environment.

We eat

We grow food on farms

We shop to buy the food

We use cars to get to the shops

These are just a very few of the things that people do.

improve the environment?

What is wrong with that?	What can be done to improve things?
Farms need space to grow more food – wildlife **habitats** destroyed	Wildlife reserves
Fertilizers and **pesticides** to grow more food – can destroy wildlife	Farm **organically** or use less harmful fertilizers and pesticides
Waste	Recycle
Too much packaging	Use **biodegradable** packaging
Too much packaging	Use less packaging
Litter	More bins
Noisy	Quieter road surfaces
Fumes – global warming and illness	Use electric or **solar-powered** cars
Dangerous roads	Speed limits
Traffic jams	Use alternative forms of transport

What a mess!

Litter looks horrible. Why do people leave their litter behind?
We all drop litter by accident. You could be going to put a crisp packet in a bin, but then the wind grabs it and blows it out of your hand.
Usually people drop litter just because they are too lazy to put it in a bin.

Often there aren't any bins, or they are full anyway.
They do need to be emptied more often, but you could go and find an empty one.

Litter doesn't matter, anyway. Someone like a street cleaner always comes along and picks it up.
That's not true! Street cleaners can't go everywhere! Anyway, you can't leave it all to them. Everyone should pick up their own litter, it only takes a few seconds. If rubbish is left lying around, it spoils places.

When I've dropped litter, I wander off somewhere else. It doesn't bother me. I don't care what I leave lying around.

What about other people? Don't you care how they feel? The litter you drop is really harmful to wildlife as well. Broken glass can cut animals' paws and start fires. Birds can be strangled by those plastic rings that hold packs of drink cans together and rats love the food you throw away.

Rubbish dropped in UK

Year	Million tonnes
1964	5
2000	25

OK, you win. I'll put this in the bin.

Clean it up!

So what about...

WASTE MINIMISATION

Your guide to the

step 1

REDUCE

The easiest way to minimise your waste is to avoid creating it in the first place!

step 2

REUSE

If you cannot avoid waste then why not REUSE it?

step 3

RECYCLE

RECYCLING can also help your Dustbin Diet

'DUSTBIN DIET'

The "Throwaway Society" is a term used to describe the 1990s lifestyle because it created so much unnecessary waste.

Problems include overpackaging and disposable products.

Reducing your waste can help to...

- save energy
- save scarce resources
- reduce litter
- conserve landfill void space

Reducing the amount of waste you create is only the first step of the Dustbin Diet - there is still a lot more you can do to starve your bin.

A lot of "waste" usually thrown away can actually be Reused, either by yourself or others.

Reuse gives items a second life (or even 3rd, 4th ...) before they are recycled or thrown away.

By reusing you will...

- save energy
- save scarce resources
- save yourself money
- conserve landfill void space

ONCE YOU HAVE REDUCED AND REUSED AS MUCH AS POSSIBLE ANYTHING ELSE MUST BE WASTE, RIGHT?

WRONG!!

Many materials can actually be Recycled - another essential element of the Dustbin Diet.

Recycling is the reprocessing of used materials to make new products. By using your local recycling facilities you will...

- save energy
- save scarce resources
- prevent valuable materials from being landfilled

WASTE MINIMISATION

Dealing with problems – looking at traffic

Traffic is an environmental problem that can be complicated to solve, with people on both sides having a good point to make.

We need to look at this kind of problem very carefully if we are to find a **solution**.

NORDON NEWS

21 March 2006

YOUR LOCAL PAPER

BOY ON BIKE HURT

By Ivor Penn

Gary Smellifeet (aged 9) was knocked off his bike in New Road last Tuesday at 5.30 pm, in yet another accident on the Lionsway Estate. Gary's mother explained he was a very careful cyclist. She added, "The only problem is the traffic – there's too much of it and it's going too fast."

Mrs Kerry White (34), who witnessed the accident, said, "The boy just skidded, right in front of the car. The driver didn't have a chance to stop. The problem is the kids – they shouldn't be allowed to cycle round here. Everyone uses New Road as a shortcut to get round the traffic jams at the lights on the main road."

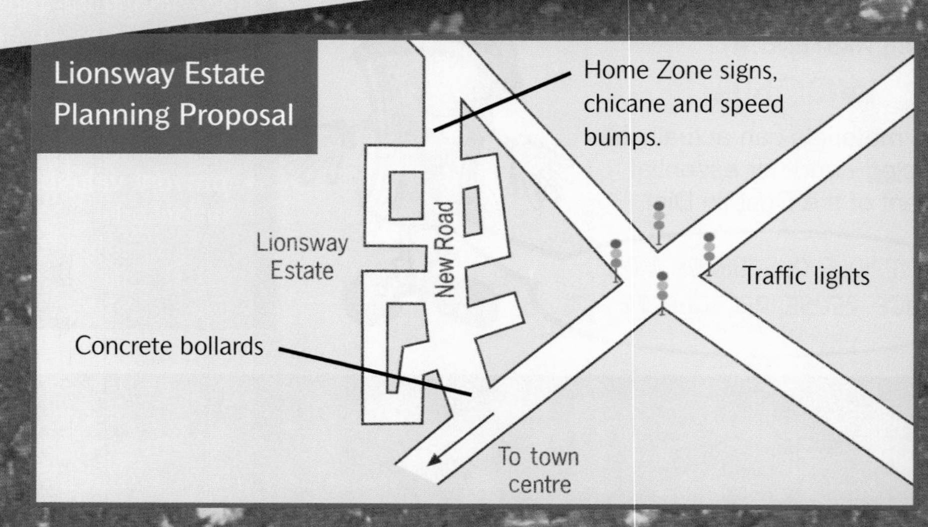

Solving the problem

You could follow the **method** Nordon Council used to solve New Road's traffic problems.

Nordon Council Traffic Report: New Road

1. Make **observations**

- Count cars using road — *9943 a day*
- Record number of accidents — *19 in three months*
- Check existing traffic controls — *30 mph speed limit*
- Observe pedestrian and cyclist use of roads — *People rarely walk or cycle*
- Design and hand out a questionnaire asking residents what they feel about the traffic. Record results. — *Hate it, dangerous, noisy and smelly*

2. Suggest a possible solution — *Stop traffic taking a short cut along New Road*

3. Design an experiment to check this is the correct solution.

Put up concrete bollards across one end of New Road, blocking it so that traffic can't get through.

4. Make observations

- Count cars — *109 cars a day*
- Record number of accidents — *One in three months*
- Observe pedestrian and cyclist use — *People walk to shops, children cycle and skateboard*
- How do residents feel now? — *Very happy*
- Any disadvantages of the new scheme? — *Longer traffic jams along main road*

5. **Interpret** information

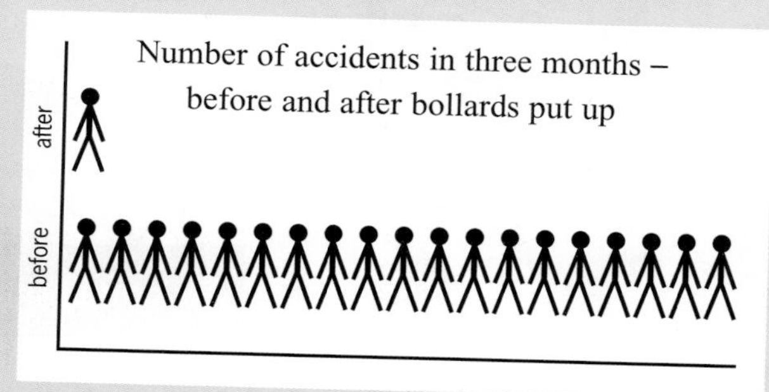

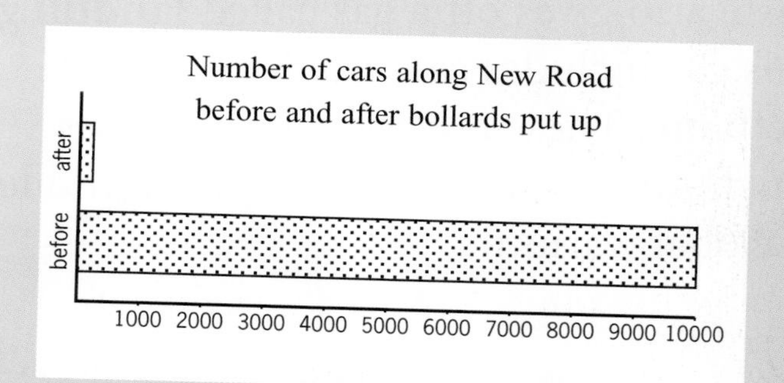

6. Draw **conclusions**

Lionsway Estate is much safer and everyone living on it is much happier. Steps are to be taken to make sure that it is recognized as a residential area with 'Home Zone' signs, chicanes and speed bumps and a 20 mph speed limit.

The Lionsway Estate is a great place to live now. I can go round to my mates without Mum making a fuss.

Now you see it, now you don't

Most waste in Britain goes to **landfills.** These are enormous holes in the ground (often old **quarries**). First the hole has to be lined with plastic and clay so that it is waterproof, because when waste is rotting it makes a liquid called **leachate.** Leachate contains poisonous chemicals that harm the environment if they escape. Rotting waste also gives off a gas called **landfill gas.** This gas includes **carbon dioxide** and **methane.** These are 'greenhouse gases' which can pollute the atmosphere and lead to global warming.

A landfill site

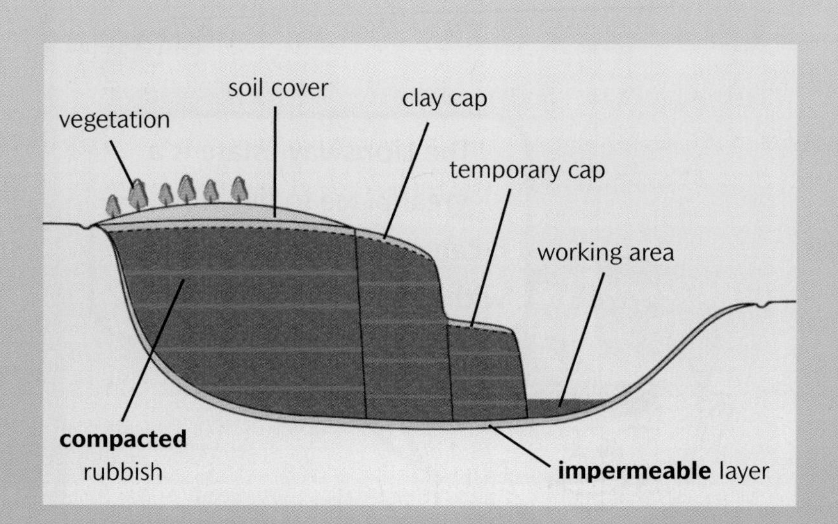

When the landfill is lined, lorries tip waste into one section of it. Next, landfill compactors (machines like bulldozers, weighing 25 tonnes) spread the waste out evenly and squash it flat. When a section is full, it is covered with earth and then another section is filled.

An incinerator at Ellesmere Port, Liverpool

When the landfill is completely filled, it is covered with a layer of clay, which is waterproof, and then with soil, so that grass, shrubs and trees can be planted. The waste takes over 30 years to rot down, and all that time the leachate and landfill gas have to be collected and disposed of. Leachate is either treated in a sewage plant, or is made safe on site. Landfill gas is usually either burnt in flares, or burnt to provide heat or to make electricity.

Many landfills in Britain are nearly full. It is difficult to find places for new ones, for two main reasons. First, landfills need to be sited where they cannot damage the environment. Second, most people do not want a landfill near where they live. Experts are therefore looking at other ways to get rid of waste. One way is to build incinerators, in which waste can be burnt.

When the waste is burnt, it makes ash, which is then taken to a landfill. The ash takes up much less space than the original waste, and is safer because it does not produce landfill gas, and there is much less leachate. The heat given off by the burning waste produces electricity and hot water. Therefore, although incinerators cost a lot to build, their impact on the environment is less than that of landfills.

This must stop!

Your dog - Message

From: BAKER, Mick
To: sophie@sophieshouse.net
Cc:
Sent: Thursday, May 10, 6:03 pm
Subject: Your dog

Hi Sophie, your dog pooed on the pavement outside our house again this morning, and Katy stepped in it. She had to go back to wash her shoe, and she walked it on the new carpet, and Mum was really mad. Mum had to wash the carpet, and Katy got poo on her hand when she was washing her shoe, and Mum made her wash her hand in antiseptic, which stung, and she started crying. We ended up late for school, and I got into trouble with Mrs Price. So tell your Dad to pick up your dog's poo, will you?
Mick

From: sophie@sophieshouse.net
To: mick@baker.net
Cc:
Sent: Thursday, 10 May, 2005 8:11 pm
Subject: dog poo

Hi Mick

I showed your email to Dad and he's promised to buy a poop-a-scoop to pick up Sheba's poo.

xxxxxx Sophie

29, Thunder Road
Yawning
YW16 9JA

Highway Maintenance
Morton Borough Council
Morton

6 June 2005

Dear Sir or Madam

Our road is very noisy. There is a lot of traffic on it, especially in the morning when people are going to work or school, and in the evening when they are coming home.

During the day big lorries use the road a lot. They are sometimes so noisy that they make our house shake, and all the windows rattle. My little brother finds it scary, because our old house didn't do that.

Another thing – the traffic sometimes wakes us up in the middle of the night. There is a pothole outside our house, and when the lorries fall in it, they make a loud CLUNK. The empty gravel lorries are especially bad. Sometimes I can't get back to sleep, and then I feel very tired at school.

Please, can you make our road quieter?

Yours faithfully

Chloe Goodchild

MORTON BOROUGH COUNCIL
HIGHWAY MAINTENANCE
MORTON
Tel: 01998 3456214 email Martina@MBC.co.uk

Miss Chloe Goodchild
29 Thunder Road
Yawning

12 June 2005

Dear Chloe

Thank you very much for your letter about Thunder Road. It is due to be resurfaced in three months' time. We are going to use a special **porous asphalt**, which reduces the noise that vehicle tyres make on road surfaces.

We have completed a survey and found that many of the lorries that use Thunder Road are on their way to the motorway. We plan to put up signs which will divert them to the motorway before they get to your village. This should help you get a good night's sleep!

Yours sincerely

Alex Martin

Highway Maintenance Manager

Recycling

What goes into our waste bins?

It is possible to recycle about 80% of our waste.

Contents, by weight, of an average bin

%	Contents
30%	kitchen and garden waste
27%	paper and card
9%	glass
6%	plastic
5%	metal
3%	textiles
20%	other (dust, ash, **crockery**, etc.)

Glass

The **consumer**

Bottles and jars should be clean.

Put glass into the correct colour bottle banks.

Doorstep collection

Reprocessing factory smashes glass into *cullet*.

New bottles and jars made.

Food factory

Supermarket

Newspapers and magazines must be clean and dry. No plastics, catalogues or phone directories.

Papers in paper bank

Papers sorted and taken by lorry or freight train to paper factory.

Paper mixed with water and chemicals – paper breaks down into pulp. More ingredients added e.g. bleach or oxygen to make paper whiter, or new wood fibre to make paper stronger.

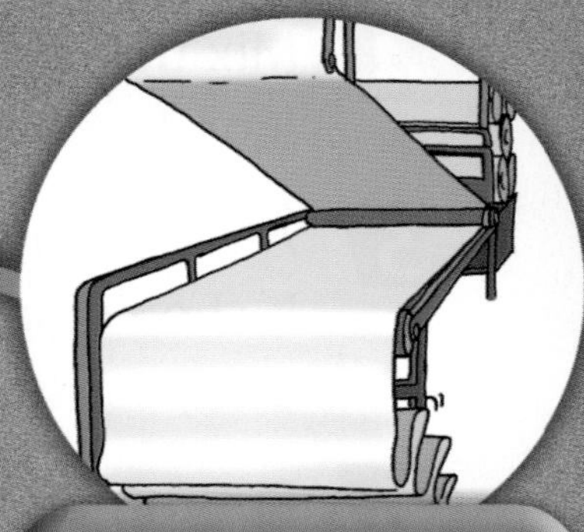

Pulp made into products – e.g. paper, card, toilet rolls.

Products taken to shop or supermarket and bought by consumer.

The **consumer**

Doorstep collection

Recycling paper

- Saves chopping down trees
- Saves energy
- Saves water

Energy and water used in making paper

Scale	
100%	non-recycled paper
60%	recycled paper
0	

40% less water and 40% less energy is used to make recycled paper

How to make compost

You will need:

- a compost bin
- a patch of bare earth
- some 'green' waste that breaks down easily:
  fruit and vegetable scraps
  contents of vacuum cleaner
  grass (from mowing lawn)
- some 'brown' waste that breaks down slowly:
  dead leaves
  crumpled newspaper
  egg shells

Method

1. Put the compost bin on bare earth.
2. Fill it gradually with layers of 'green' and 'brown' waste.
3. Water the bin if the contents look dry.
4. Mix the contents of the bin every 3 or 4 weeks.
5. After several months, inspect the mixture. It should have changed into a dark brown compost smelling of clean earth.
6. Put this on the garden to help plants grow well.

Add other things that rot down easily, such as dirty sawdust or straw from the hutches of vegetarian pets, e.g. rabbits, hamsters and guinea pigs.

If you don't have enough space for a compost bin, make compost in a worm bin.

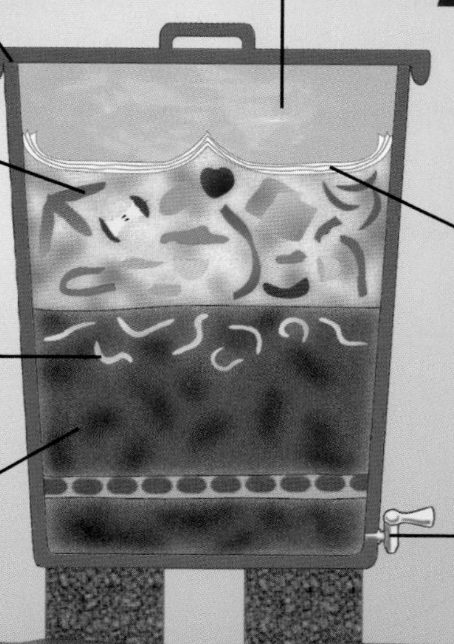

ACE ADVERTISER

6th November 2003

NEWS AND VIEWS

NEW IDEAS, NEW INDUSTRIES

Recycling is here to stay. There's no way we can get away from that. Most of us go to recycling centres and push bottles in the bank and listen to them smash. We shove newspapers through the slot – but what happens then?

By Richa Kidd

Never before have there been so many old newspapers and glass bottles to deal with. These are new **raw materials** for new industries, and people are busy coming up with exciting new ideas about how to use them.

Take glass, for example. Your old Marmite jar (and any other glass) may go into making glasphalt, a new substance used in making roads, which contains 30% crushed recycled glass. Go to Camden Town Hall, in London, and you can walk over broken glass in your bare feet – some floor tiles there are made with 80% recycled glass!

Newspapers aren't just used for making recycled paper and card. Someone has had the bright idea of turning the fibre from newspaper into good **insulation** for homes, to keep them warm.

So keep your eyes open, and your imagination working. Maybe you can come up with a new idea about what to do with recycled waste, and make your fortune from it.

Air pollution

Industrialization

Two hundred and fifty years ago most people in Britain lived in small towns or villages. They worked in or near their homes. They spun and wove wool by hand, and sewed clothes by hand. They walked or rode wherever they wanted to go. They had fires, often of wood or peat, to cook on and keep them warm. Windmills and water-wheels were used to drive machinery.

In 1782, James Watt invented a steam engine which could drive machines much more reliably. More and bigger machines were developed, and more and more factories were built. People moved near the factories to find work. Many of the factories were in towns like Manchester and Leeds. With the growing population these towns quickly grew into big cities.

Transport developed too. Canals were built to carry raw materials to the factories and factory goods to the consumers.

In 1829, George Stephenson's steam locomotive 'Rocket' reached the great speed of 21 mph! This was very fast to people who were used only to horse power. By 1850 trains were going at 60 mph! Coal, a very smoky fuel, was used to power steam engines and locomotives and to heat homes.

James Watt and the first steam engine

By 1900, thousands of factories across Britain were belching out filthy smoke from their chimneys. Many more people now lived in the towns and cities, and the dark smoke from the open fires in their homes added to the dirty air.

By the 1950s, coal-fired power stations, which made electricity, were also puffing out stinking smoke. In London, in 1952, the air became so thick with pollution that the city was swallowed up in the Great London Smog – a fog so thick that people got lost just trying to cross the road.

The Great London Smog

The government had to do something. In 1956, it passed the Clean Air Act which stopped people in some areas from burning coal. Instead, people burned smokeless fuel, or turned to gas or electric heating, which is much cleaner.

After the Act, air pollution from factories and coal fires grew much less. However, more and more people began to buy cars, finding them more convenient than public transport.

Also, whereas once barges and freight trains carried most of the country's freight, the better road network meant that lorries now could, leading to the decline of canal and train transport.

Today, most air pollution in Britain comes from road traffic. The pollution is unhealthy for all of us, but especially for people with weak hearts or lungs. If people have asthma, air pollution can trigger an attack.

Some countries have worse air pollution than Britain. New Delhi, capital city of India, is one of the most polluted cities in the world.

- In 1997, traffic fumes made up 65% of New Delhi's air pollution.
- In 1995, 9859 people were thought to have died because of air pollution.

Let's prevent air pollution caused by traffic!

Make laws
- All new cars to have catalytic converters – make exhaust fumes cleaner
- Speed limits

Improve car design
- Electric cars
- **Hybrid** engines – electric and petrol together
- **Fuel cells**
- Car shape can cut fuel used

Everyone can help
- Walk or cycle on short journeys
- Share cars (but only with friends)
- Take public transport, e.g. bus or train

Up in the air

Global warming - Message

From: Dev@inspcs.co.uk
To: meena@acecol.ac.uk
Cc:
Sent: Wednesday, May 21, 2005 6.31 pm
Subject: Global warming – is our house going to be flooded?

Dear Meena

The TV news said that global warming means the sea level will rise, and London and lots of other cities round the world will be flooded. Are we going to be all right?

Best wishes

Dev

Dev@inspcs.co.uk

From: Meena@acecol.ac.uk
To: Dev@inspcs.co.uk
Subject: Global warming

Hi Dev –

There's a layer of air all around Earth, called the atmosphere. Gravity holds the atmosphere on to Earth, just as gravity holds us to the ground. That means air can't float off into space.

Earth gets light from the Sun. You can feel its warmth when you stand outside in the sunshine. What the atmosphere does is to protect the Earth from the full fierceness of the Sun's heat during the day, and keep the warmth in at night. The atmosphere also shields Earth against harmful radiation from the Sun.

XX Meena

Global warming - Message

From: Dev@inspcs.co.uk
To: meena@acecol.ac.uk
Cc:
Sent: Wednesday, May 21, 2005 6.38 pm
Subject: Global warming

Meena, I asked about global warming!

Dev@inspcs.co.uk

From: Meena@acecol.ac.uk
To: Dev@inspcs.co.uk
Subject: Global warming

Give me time to explain, Dev. I'll do a drawing.

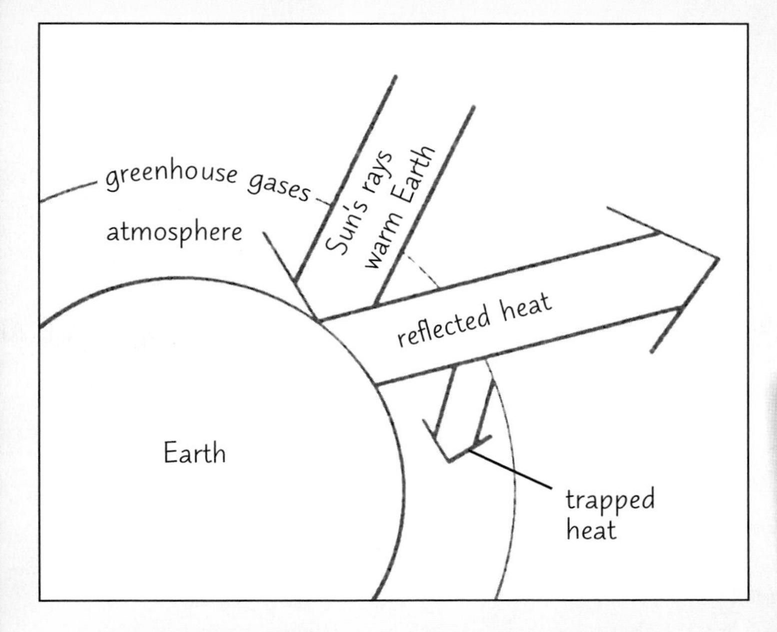

You know air is made of different gases, don't you, Dev? Some gases in the atmosphere (such as **carbon dioxide** and **methane**), let sunlight through to the surface of Earth, but when the infra-red heat rises up from Earth and hits them, the gases send the heat back to Earth. Gases that do this are called greenhouse gases, because they keep Earth warm, like the glass of a greenhouse keeps the air inside it warm.

There is more greenhouse gas in the atmosphere now than there used to be. Earth is becoming warmer, and many scientists think that this is because of the increase in greenhouse gas. They say that some of the ice in the Arctic and Antarctic is melting, and so sea levels are rising. They say Earth's climate is changing, and that there will be floods and droughts all over the world. They think maybe pollution is causing the problem.

Global warming - Message

From: Dev@inspcs.co.uk
To: meena@acecol.ac.uk
Cc:
Sent: Wednesday, May 21, 2005 7.23 pm
Subject: Global warming

Couldn't Earth's climate be changing naturally? Britain was covered in ice 10,000 years ago, in the Ice Age. There wasn't any industry then, but still Earth warmed up.
Dev

Dev@inspcs.co.uk

From: Meena@acecol.ac.uk
To: Dev@inspcs.co.uk
Subject: Global warming

That's true. But we make a lot of greenhouse gas through traffic, old-fashioned industry and fires. Changing the way we live, by using **renewable energy**, will make Earth cleaner. We mustn't risk damaging Earth's climate.
XXX Meena

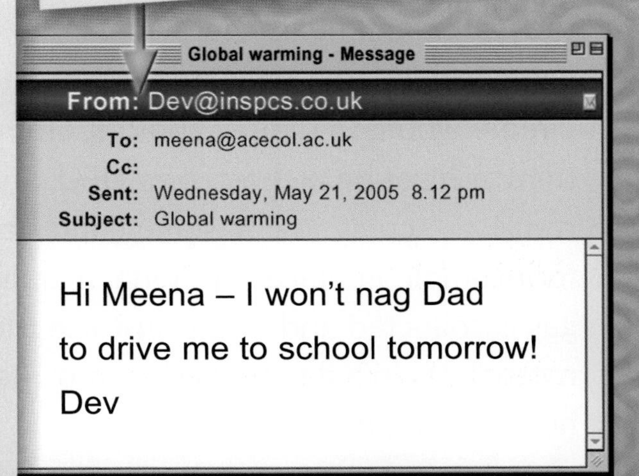

Global warming - Message

From: Dev@inspcs.co.uk
To: meena@acecol.ac.uk
Cc:
Sent: Wednesday, May 21, 2005 8.12 pm
Subject: Global warming

Hi Meena – I won't nag Dad to drive me to school tomorrow!
Dev

People are not the only problem

Marsh gas

When marsh plants die, they rot under the water, and as they rot they produce **methane**. This is a colourless gas that has no smell. Because it is lighter than air, it collects just above the surface of the water, and floats there, making a pale glow. It looks very creepy, and people who don't know about it often think it must be a ghost, or a UFO, and run away screaming! In some countries, for example in places in China, marsh gas is collected and burnt (just like the natural gas in a gas cooker) to provide heating for cooking and homes.

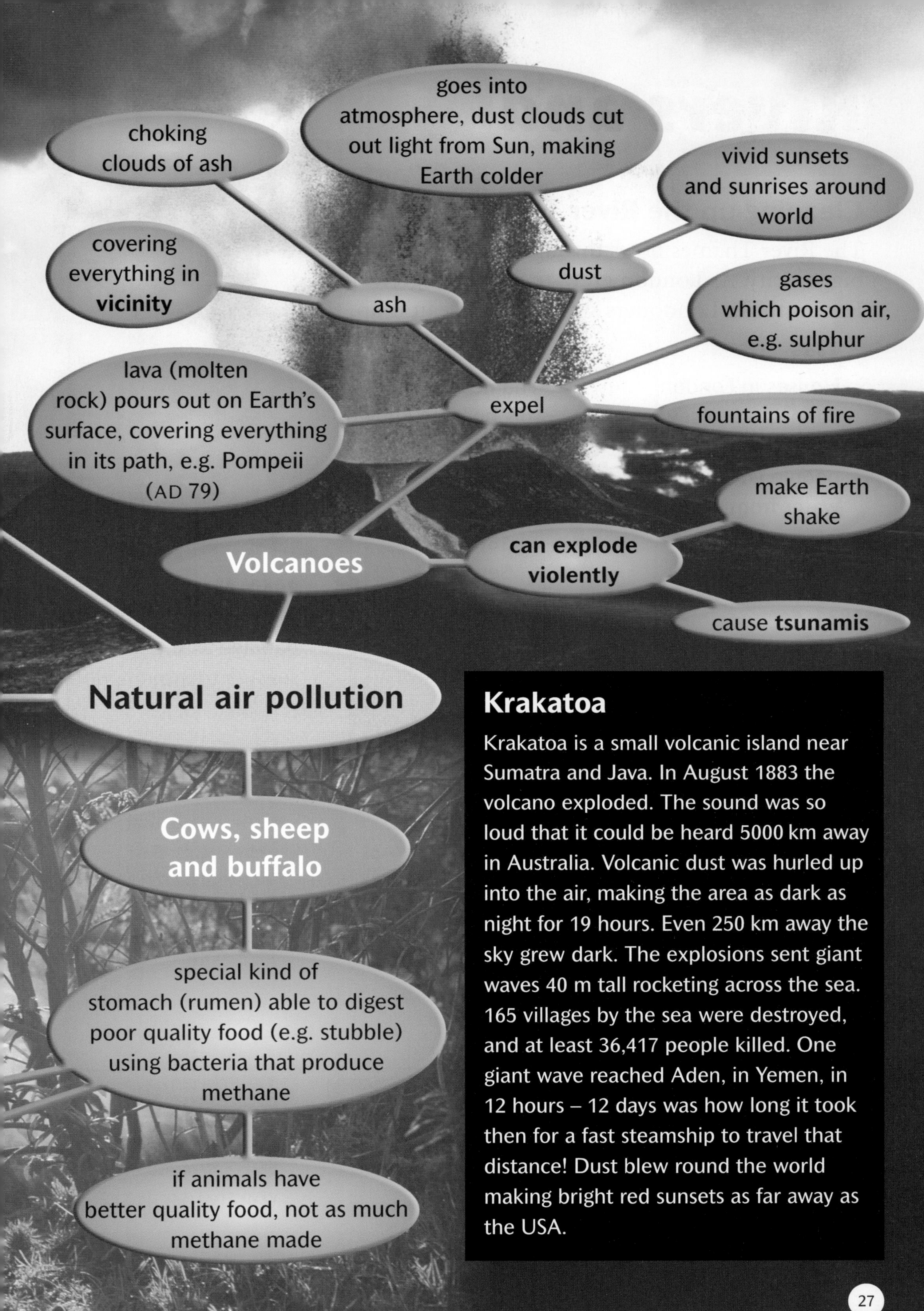

choking clouds of ash

covering everything in **vicinity**

goes into atmosphere, dust clouds cut out light from Sun, making Earth colder

vivid sunsets and sunrises around world

ash

dust

gases which poison air, e.g. sulphur

lava (molten rock) pours out on Earth's surface, covering everything in its path, e.g. Pompeii (AD 79)

expel

fountains of fire

make Earth shake

Volcanoes

can explode violently

cause **tsunamis**

Natural air pollution

Cows, sheep and buffalo

special kind of stomach (rumen) able to digest poor quality food (e.g. stubble) using bacteria that produce methane

if animals have better quality food, not as much methane made

Krakatoa

Krakatoa is a small volcanic island near Sumatra and Java. In August 1883 the volcano exploded. The sound was so loud that it could be heard 5000 km away in Australia. Volcanic dust was hurled up into the air, making the area as dark as night for 19 hours. Even 250 km away the sky grew dark. The explosions sent giant waves 40 m tall rocketing across the sea. 165 villages by the sea were destroyed, and at least 36,417 people killed. One giant wave reached Aden, in Yemen, in 12 hours – 12 days was how long it took then for a fast steamship to travel that distance! Dust blew round the world making bright red sunsets as far away as the USA.

Rivers and beaches

Cleaning up the River Thames

The River Thames flows through London, Britain's biggest city. As London grew, the Thames grew dirtier. For centuries Londoners struggled to clean it up.

1831–66
Four **cholera epidemics** – over 35,000 people died.

1833
Houses in London connected to **sewers** going into Thames – people got drinking water from river.

1858
Thames so smelly this became known as 'Year of the Great Stink'.

1882
Sewage treated for first time.

1940–45
Second World War – bomb damage to sewers and sewage treatment works.

Oil spill alert!

1996

15 February
Oil tanker *Sea Empress*, loaded with 131,000 tonnes of crude oil, ran aground at entrance to port of Milford Haven – oil spilt.

16 February

16–24 February
Planes sprayed oil on sea surface with dispersant.

16–27 February
950 people involved in cleaning spilt oil from beaches.

16 February–2 March
Over 20 ships tried to recover oil remaining at sea.

17 February
High wind and fierce tides swept *Sea Empress* on to rocks – more oil spilt.

1950s
River badly polluted by sewage, industry, power stations and gas works. There were no fish between Fulham and Tilbury.

1960s
Sewage treatment works improved – industrial pollution controlled – 1974 first salmon in Thames since 1833! Salmon need clean rivers. Since 1974, 115 different kinds of fish seen.

1989 onwards
Sometimes Thames low in oxygen – 1989 boat 'Thames Bubbler' built – can inject up to 30 tonnes oxygen a day into river.

1997

21 February
Sea Empress towed by tugs to harbour where 59,000 tonnes of oil pumped out. 72,000 tonnes of oil spilt at sea.

22 February–2 March
3100 oily birds cleaned at emergency centre. 7000 oily birds washed ashore, number killed much higher.

5–8 April
Main tourist beaches clean.

Late April
Not all birds cleaned of oil survived when released. Only 3% of guillemots survived.

June
All beaches in area clean, even the ones that had not been given special treatment.

Land use

How do we use the land?

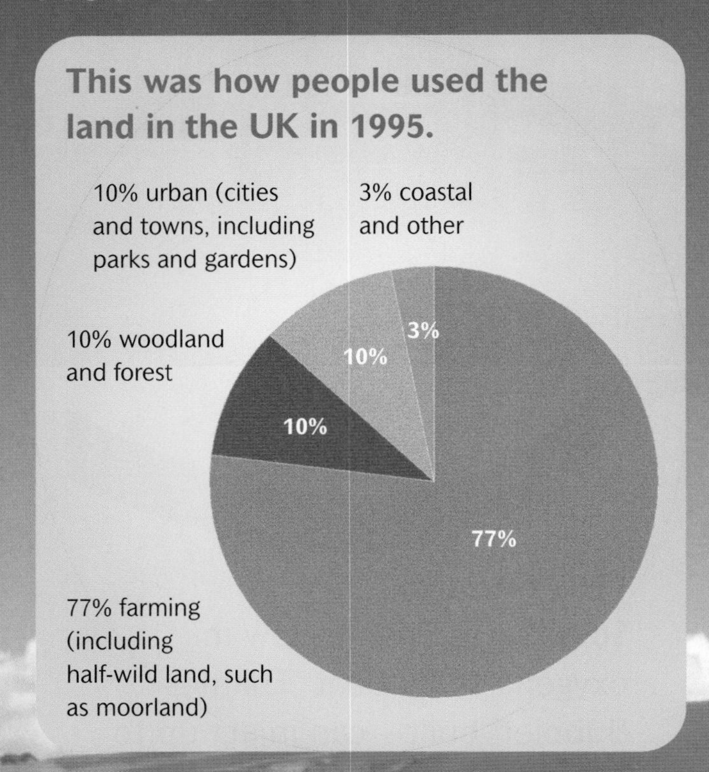

People very often change the way they use land. New technology and new ideas spark off changes in the way we want to live, and the way we use land changes to match our requirements. For example, as more people want to fly by plane, green fields are turned into new airport runways. Changes in land use can affect people and their environment (for instance, a new airport leads to a big increase in traffic). Therefore the government has made laws which say that people who want to build a new building, road or runway, must get **planning permission** first.

For a long time, few people worried about how changes in land use were affecting wildlife. Partly because of the changes, over 100 species in the UK were lost in the last century, including 7% of dragonflies and 5% of butterflies. In the last 50 years, about 97% of **meadows** rich in wild flowers have been ploughed up for farming, or built on, destroying insect habitats. This means insects have fewer places to live, increasing their chance of becoming extinct. The last orange-spotted emerald dragonflies died after spilt sewage polluted the only breeding ground they had left.

Therefore, people should think about wildlife as well as themselves when they change the way they use land. The government now offers farmers grants of money to encourage them to create wildlife habitats on their land.

Why and how do we change the way we use land?

More people

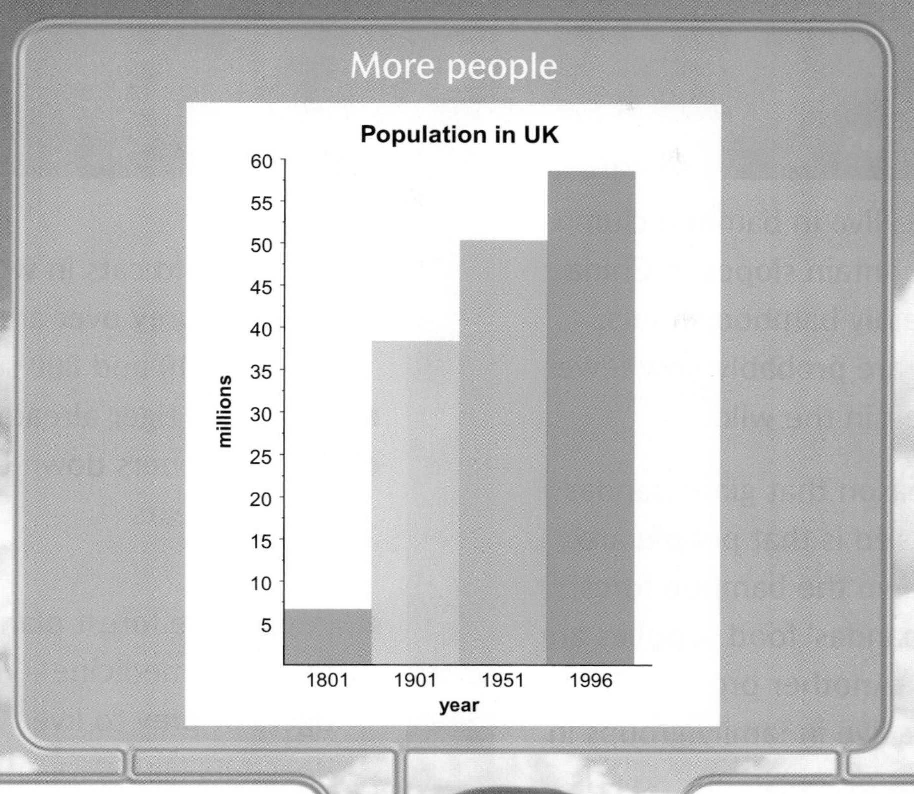

Need to get to work

Need somewhere to live

Need to eat

More cars

Changes in farming to produce more food cheaply

More roads

Problems, e.g. 'mad cow' disease

Create pollution

New towns and housing developments

Chemical fertilizers and pesticides

Big fields and no hedges – bad for wildlife

Big supermarkets, out-of-town shopping centres

Can create pollution

Animals in danger

Giant pandas

Giant pandas live in bamboo clumps on steep mountain slopes in China. They eat mainly bamboo shoots. Sadly, there are probably now fewer than 1000 left in the wild.

The main reason that giant pandas are endangered is that people are chopping down the bamboo forests, and so the pandas' food supplies are running out. Another problem is that giant pandas live in family groups in small patches of forest, and so it is difficult for them to meet other pandas, to breed and have cubs. A further danger comes from **poachers**, who kill or capture the giant pandas.

In order to save the giant pandas, **conservationists** need to study them in order to find out their exact numbers, and how they live. This will help us know what they need to survive. One thing that can be done is to plant special strips of bamboo forest (called panda-corridors). These link one patch of forest with another so that pandas can meet more easily. Also, more panda reserves can be set up (there are already some in China). In the reserves, trained staff protect and look after the pandas.

Tigers

Facts

- Largest wild cats in world
- Hunt for prey over area of between 10 and 800 sq km
- 3 types of tiger already extinct
- Tiger numbers down by 95% in last 100 years

Dangers

- People use forest plants for fuel, food and medicine – nowhere for tigers or prey to live
- Poachers kill tigers and tiger prey
- People use tiger bones in traditional Chinese medicine
- People kill tigers to stop them eating their animals

Actions

- Protect blocks of forest – plant forest corridors between them
- Ban medicinal use of tiger bones
- Train forest guards – give them vehicles to combat poachers
- Explain to local people why it is important to conserve tigers

African elephants

Facts

- Live up to 70 years in groups of 6–70
- 1979 more than 1.3 million elephants
- 1989 fewer than 544,000

Dangers

- 1979–89 poachers killed over half of Africa's elephants for ivory tusks
- Many more people in Africa – need to use elephants' habitat

Actions

- Make safe game reserves. Kruger National Park very secure – elephant numbers growing quickly
- Tell local people elephants useful – bring in tourists
- 1990 ban put on international trade in ivory
- Use alternative materials to ivory

Otters (in England and Wales)

Facts

- Live in holes in river bank or at foot of tree
- Eat mainly fish and water birds, also frogs in spring
- Sleep in daytime
- Widespread in 1950s – 20 years later had almost disappeared

Dangers

- Loss of habitat – wet land drained, fewer plants on river banks
- Poisoned by a pesticide widely used in 1960s
- Increased road traffic – at least 300 run over since 1988

Actions

- Plant more plants along river banks
- Keep people away from **protected areas**
- Clean up rivers. (This works well – otters seen in city centres of Birmingham and Newcastle-upon-Tyne)
- Build otter walkways under new roads

International and national action

September 2002

Steps to improve environment: 187 NATIONS AGREE

The United Nations Earth Summit in Johannesburg, South Africa, hasn't changed our world overnight. But steps have been taken which will, slowly, begin to improve our environment. World leaders have promised to make and use chemicals only in ways that won't harm human health and the environment as much. They have agreed that more electricity should be generated using the power of the Sun, wind and tides.

August 2002

WWF IN AMAZON

By R G Branwell

The Amazon, in South America, is the largest river basin in the world. World Wide Fund has been working hard, together with local governments and people, to keep it safe for the children of the future.

With the support of WWF and the World Bank, the government of Brazil has said it will make 25 million hectares in the Amazon into national parks and reserves. WWF has also helped set up the Manu National Park in Peru, which protects nearly 10% of known bird species.

WWF has set up a factory to give some local people work so that they do not need to chop down the rainforest.

WWF has projects all over the world, including in Britain. Its success proves how essential it is to have the support of local people in order to protect the environment.

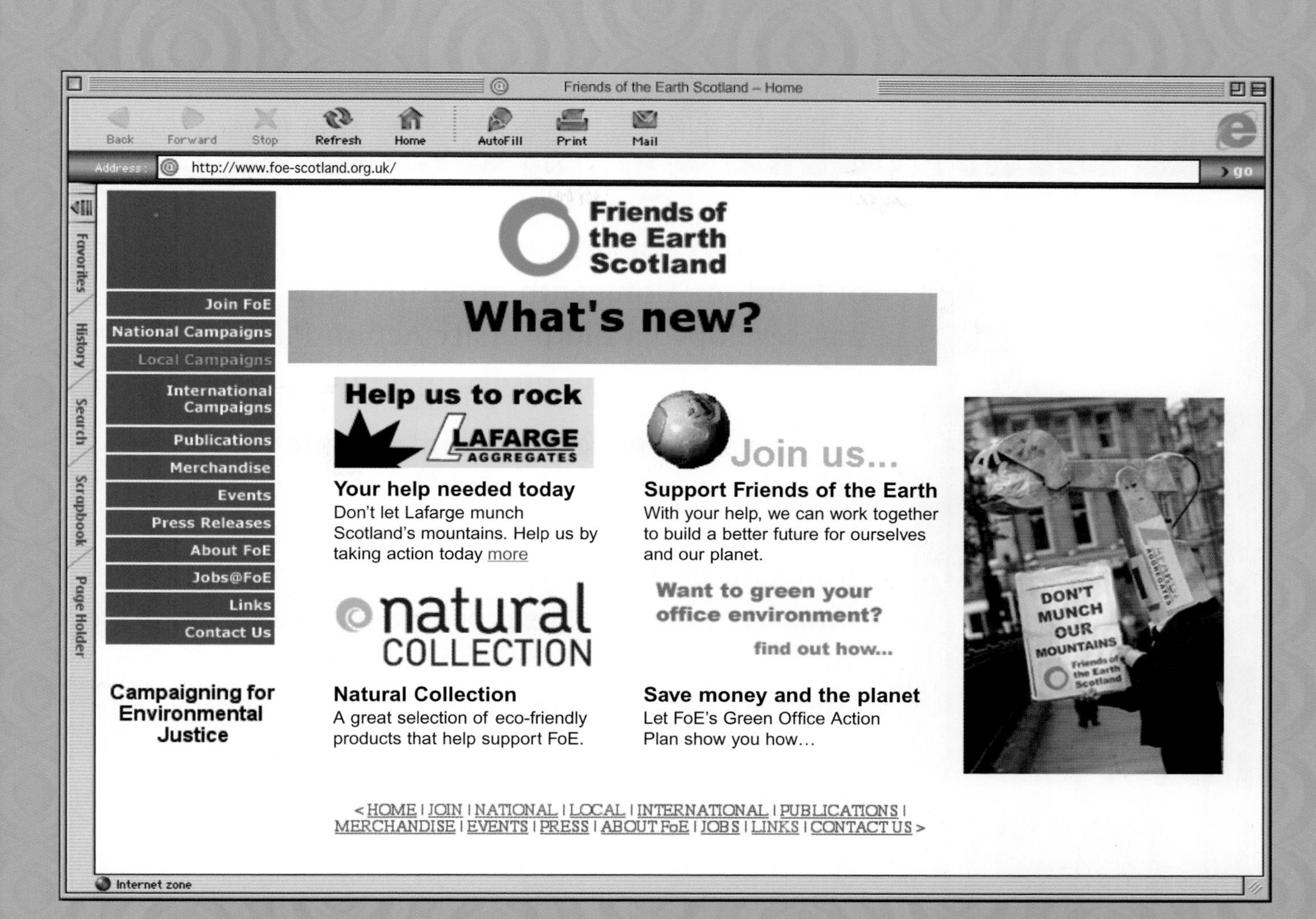
Friends of the Earth Scotland – Home
Back
Forward
Stop
Refresh
Home
AutoFill
Print
Mail
Address:
http://www.foe-scotland.org.uk/
go
Favorites
History
Search
Scrapbook
Page Holder
Friends of the Earth Scotland
What's new?
Join FoE
National Campaigns
Local Campaigns
International Campaigns
Publications
Merchandise
Events
Press Releases
About FoE
Jobs@FoE
Links
Contact Us
Campaigning for Environmental Justice
Help us to rock
LAFARGE AGGREGATES
Your help needed today
Don't let Lafarge munch Scotland's mountains. Help us by taking action today more
Join us...
Support Friends of the Earth
With your help, we can work together to build a better future for ourselves and our planet.
natural COLLECTION
Natural Collection
A great selection of eco-friendly products that help support FoE.
Want to green your office environment?
find out how...
Save money and the planet
Let FoE's Green Office Action Plan show you how…
DON'T MUNCH OUR MOUNTAINS
< HOME | JOIN | NATIONAL | LOCAL | INTERNATIONAL | PUBLICATIONS | MERCHANDISE | EVENTS | PRESS | ABOUT FoE | JOBS | LINKS | CONTACT US >
Internet zone

About us
Back
Forward
Stop
Refresh
Home
AutoFill
Print
Mail
Address:
http://www.greenpeace.org/aboutus/
go
GREENPEACE
websites
English
Search
home
help
Join
about us
latest news
multimedia
press centre
campaigns
Highlights
Our mission
History
Victories
Jobs
Join
Volunteers
Ships
Contact us
Annual report
Copyright
Privacy policy
Site credits
Greenpeace worldwide
Frequently asked questions
Greenpeace science unit
About us
Greenpeace exists because this fragile earth deserves a voice. It needs solutions. It needs change. It needs action.
Greenpeace is a non-profit organisation, with a presence in 40 countries across Europe, the Americas, Asia and the Pacific.
As a global organisation, Greenpeace focuses on the the most crucial worldwide threats to our planet's biodiversity and environment.
We campaign to:
--Stop climate change
--Protect ancient forests
--Save the oceans
--Stop whaling
--Say no to genetic engineering
--Stop the nuclear threat
--Eliminate toxic chemicals
--Encourage sustainable trade
And we believe that the struggle to preserve the future of our planet is not about us. It's about you. Greenpeace speaks for 2.8 million supporters worldwide, and encourages many millions more than that to take action every day.
Greenpeace oil campaigner Paul Horsman at site of oilspill, near Usinsk, Russia.
Privacy
Copyright
Contact us
Sitemap
Help
Home
Internet zone

Community action

Chelmsford is a large town about 45 km north-east of London. Along with other towns in the UK, it has developed a Biodiversity Action Plan – a list of things to do to help protect wildlife and habitats.

Towns can provide important wildlife habitats, in school grounds, gardens, parks, ponds, rivers and waste land. The Action Plan helps wildlife directly, for example, by creating local nature reserves. It also aims to encourage local people to take more interest in the wildlife near their homes and to understand and care for it. Some people may choose to volunteer to help with projects to improve wildlife habitats.

It is hoped that, when the Action Plan comes to an end, local people will continue to care for wildlife in their area.

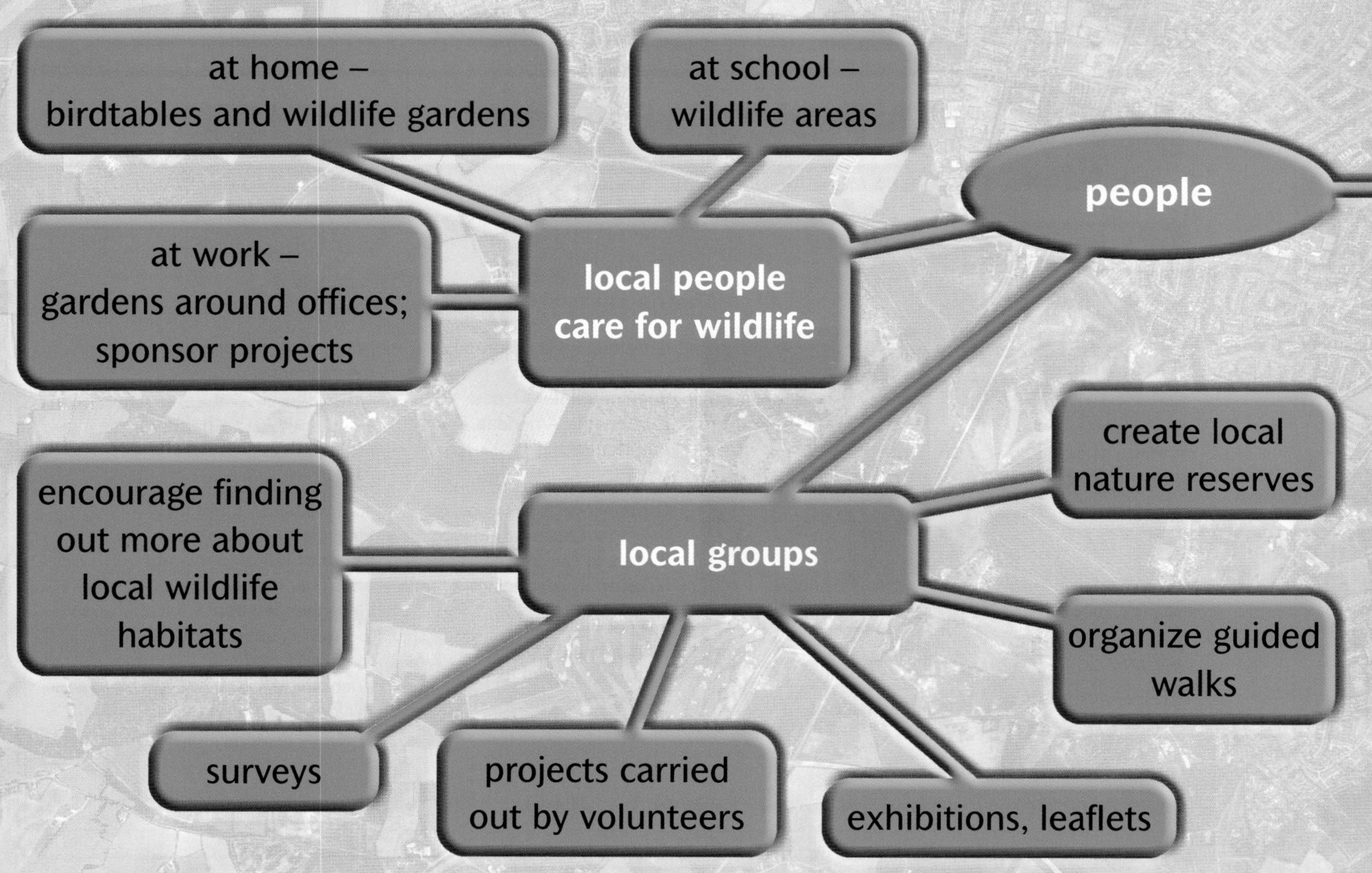

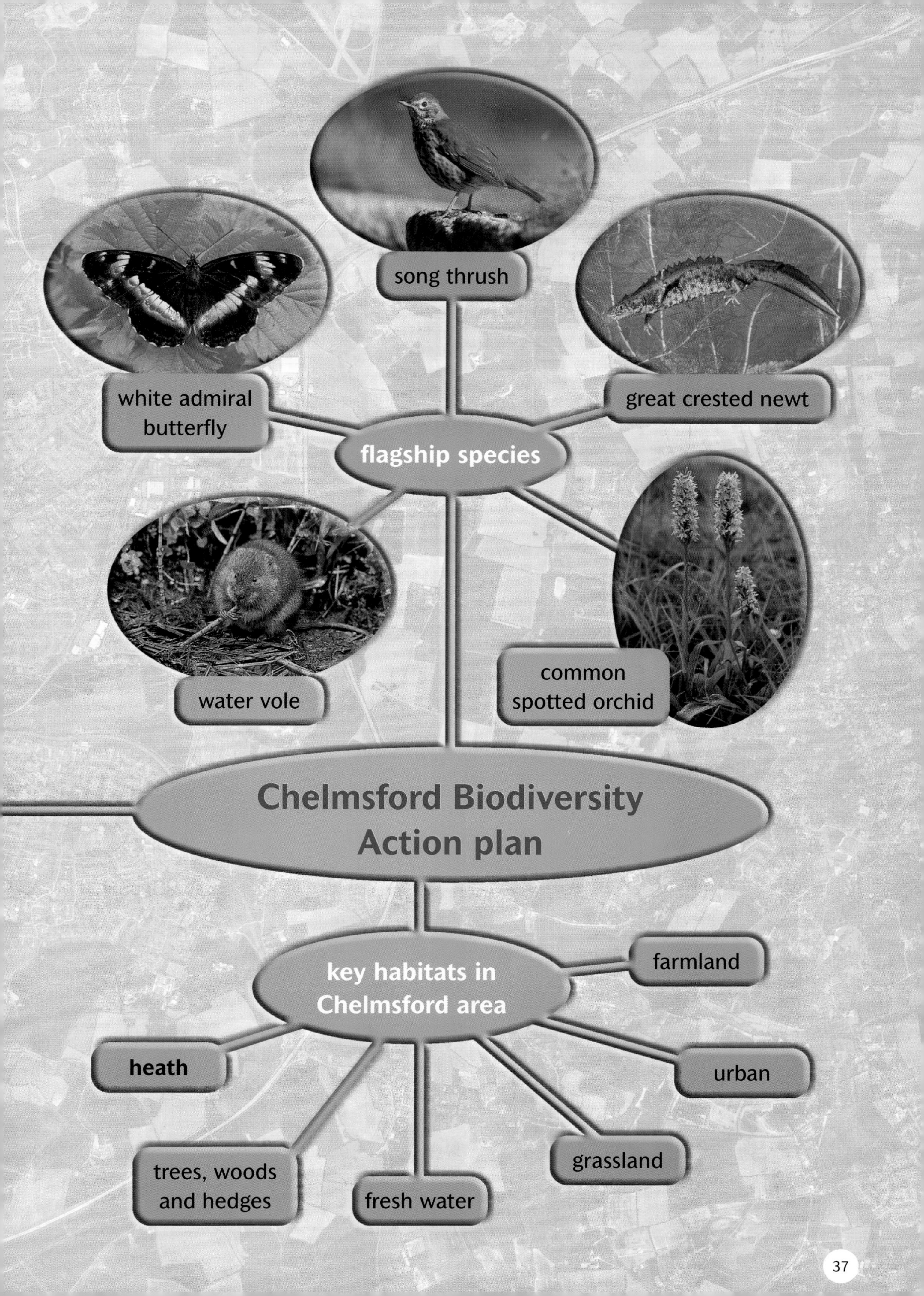
song thrush
white admiral butterfly
great crested newt
flagship species
water vole
common spotted orchid
Chelmsford Biodiversity Action plan
key habitats in Chelmsford area
farmland
heath
urban
grassland
trees, woods and hedges
fresh water

Building a wildlife garden

10
TOOLS

COMPO

9
COMPO

8
COMPO

7
ivy and honeysuckle for birds
nettles for caterpillars
buddleia for butterflies

5
curled pondweed
marsh marigold
water forget-me-not

6
for safety 75 mm squares –
space for small animals and plants

Stop press

THE PANDA TRUST
www.panda-trust.org.uk
01388 517003. e: info@panda-trust.org.uk.

Big planet, small turtle.

Needs help!

In a big world full of hungry mouths, this tiny baby turtle's life will be a struggle for survival. But marine turtles are survivors and have been travelling our vast oceans for millions of years.

Tragically, due to man's activities, marine turtles are now endangered. They urgently need our help.

The Marine Conservation Society is working to support vital turtle conservation projects around the world.

The ideal gift for someone special this Christmas.

Tel: 01989 566017 Fax: 01989 567815
www.mcsuk.org

The Marine Conservation Society is the UK charity dedicated to the protection of the marine environment and its wildlife.

MARINE CONSERVATION SOCIETY
Supported by
C&G Cheltenham & Gloucester

Picture: Turtle hatchling © Ben Osborne

www.dianfossey.org

...for someone special

Amy is a six year old mountain gorilla. One of only 650 left in the world. For just £24 a year you can adopt her and help the Dian Fossey Gorilla Fund protect all the endangered mountain gorillas from extinction. In return we'll send you a beautifully illustrated T-shirt, a photograph of Amy and your own adoption certificate.

Adoptions make a perfect gift

Call today to order your Amy adoption gift pack

Amy Hotline: 0870 241 0643

The Dian Fossey Gorilla Fund
Reg Charity No. 801160

Help save farm wildlife!

Support The Wildlife Trusts' own brand of milk, **White & Wild**, and help us to bring back a countryside rich in wildlife – from brown hares to skylarks, and water voles to wild flowers

White & Wild milk helps farmers to encourage wildlife to thrive on their farms, supports rural communities, and funds vital conservation work of The Wildlife Trusts

Please give farm wildlife a future and make **White & Wild** a success, by following these three easy steps:

1. Buy **White & Wild** milk*
2. Ask your supermarket to stock **White & Wild**
3. Spread the word!

Thank you

*log on to www.wildlifetrusts.org for further information on The Wildlife Trusts and on www.whiteandwild.co.uk for stockists and details on White & Wild

THE wildlife TRUSTS

Contribute to Conservation and have the Experience of a life time in Africa

Placements are available for enthusiastic, self-funding, volunteers to spend up to 3 months on Game and Nature Reserves in Southern Africa. Conservation work may include:

- Game Capture and Relocation (including Darting Elephants and Rhino).
- Wildlife Rehabilitation and Endangered Species Breeding Projects.
- Research & Monitoring Projects involving mammals, birds, reptiles, insects and plants.

Work on the Wild Side

AFRICAN CONSERVATION EXPERIENCE

www.afconservex.com

VINE HOUSE FARM

bird foods

Buy your bird food direct from the farm

Nicholas Watts brings you wild bird seed, grown on his award winning farm, direct to your door.

Nicholas' lifelong love of wildlife prompted him to adapt his farming methods to encourage birds back to the farm. Realising he could grow the seed that attracts and nourishes a wide variety of birds he is now able to offer you the same opportunity in your own garden.

Vine House Farm Bird Foods aims to provide the best quality home grown seed mixed specifically for wild birds at very reasonable prices.

Products and Prices

Delivery included to most parts of mainland Britain

Product	12.5kg	25kg
Black Sunflower Seeds	12.5kg – £12	25kg – £18
Sunflower Hearts	12.5kg – £19	25kg – £33
Premium Peanuts	12.5kg – £15	25kg – £24
Peanut Granules	12.5kg – £16	25kg – £26
Wheat	12.5kg – £9	25kg – £12
Rolled Naked Oats	12.5kg – £11	25kg – £16
Oil Seed Rape	12.5kg – £10	25kg – £14
Linseed	12.5kg – £5	25kg – £8

Special Mixes: The best seller!

Product	12.5kg	25kg
Mixed Seeds	12.5kg – £12	25kg – £18
Premium Mix	12.5kg – £14	25kg – £22
Finch Mix	12.5kg – £13	25kg – £20

Nicholas Watts
Vine House Farm,
Deeping St Nicholas, Spalding,
Lincs PE11 3DG

Phone: 01775 630 208
Fax: 01775 630 244
E-mail: p.n.watts@farming.co.uk
www.vinehousefarmbirdfoods.co.uk

Contact us to order or for a full price list on

01775 630 208

Vine House Farm – farming for wildlife

The environment is for everyone...

… but that doesn't mean everyone can do what they want

The boy on the skateboard wants to jump down the steps. The little children want to climb up the steps. What will happen?

If the boy goes on he will knock the children over. Then not only will the children be hurt, but they will lie in his way, crying, and he won't be able to skateboard at all. So he waits until the children are out of his way, and then does his jump – or else he skateboards somewhere else.

Change

Cars are designed to travel from one place to another quickly and easily. However, there are very often traffic jams going into towns and cities. In a traffic jam nobody goes anywhere, or they just move very slowly. Pollution comes out of exhaust pipes: gases and dust that can make people ill, and carbon dioxide which contributes to global warming. Are drivers aware of what they are doing? It is as if the boy had decided to knock the little children over, but instead of the children, it is the environment that is hurt and crying.

We all share the environment, and sometimes we need to give up, or change, things that we want to do, in order for the environment to survive. We need to come up with new, bright and exciting ways of doing things better. Can you think of a better form of transport to take the place of cars in towns and cities?

What now?

Earth is a small, beautiful planet circling in vast, uninhabitable space. It is the only place in the Universe that we know of where humans can live. But it is being damaged by what we do – by the way we are using up its resources, by the way we are polluting the atmosphere, by the way we are destroying wildlife.

We do not completely understand the way that Earth's climate works. But we know enough to tell us that we need to take more care. We need to find out more about our environment, so that we can work out how to look after it better. Then it can be a good place for everyone.

We can all do something now to help improve our environment. It may be no more than walking to school, if we live close and it's safe to do so. It may be no more than picking up litter, or recycling a drinks can, or feeding a bird. But everything helps!

Glossary

atmosphere the air around the Earth

biodegradable made of materials which rot down naturally

carbon dioxide a colourless gas that doesn't smell. People and animals breathe it out; plants absorb it.

chicane a barrier designed to slow traffic

cholera a serious disease caused by drinking dirty water

conclusion something that you decide after considering all the facts

conservationist someone who protects and takes care of the environment

consumer somebody who buys and uses a product

crockery plates, mugs, bowls, etc. made of pottery

cullet broken glass used in recycling

epidemic a disease affecting many people at the same time

fertilizer a substance added to the ground to make plants grow better

fuel cell a clean, efficient means of making electricity

habitat a place where an animal or plant usually lives or grows

heath wild area with low-lying bushes that isn't farmed

hybrid made by putting two different things together

incinerator a building where waste can be burned

insulation a substance that stops heat, or sound, going through itself

interpret to explain

landfill a place where waste is buried

landfill gas a gas given off by rotting waste in a landfill

meadow a grassy field

methane gas given off from rotting vegetable matter

method a way of doing something

observation the looking at and writing down of something that is happening

organically without using chemical fertilizers or pesticides

pesticide a substance that kills harmful insects

planning permission permission given by local or national government to allow someone to build or change a building, road or runway

poacher someone who catches or kills wild animals illegally

porous asphalt a type of road surface with holes which allows surface water to drain away and lessens tyre noise

protected area an area where an animal is kept safe from being hunted

quarry a place where stone and gravel are dug out of the ground

raw materials the basic substances from which products are made

renewable energy energy that will not run out, e.g. solar, wind and tidal

reprocessing to make used material into a new material

sewer a pipe that carries dirty water away from buildings

solar-powered using the energy of the Sun

solution an answer to a problem

tsunami a huge sea wave caused by an earthquake or a volcano

vicinity an area surrounding a place

Bibliography

Books

Non-fiction

Bailey, D. *What Can We Do About Recycling Rubbish*
ISBN: 07496 04077

Featherstone, J. *Earth Alert! Energy*
ISBN: 07502 33303

Graham, I. *Energy Forever?: Solar Power*
ISBN: 07502 37694

Harlow, R. and Morgan, S. *Pollution and Waste*
ISBN: 07534 55056

Harte Smith, A. *Food Safety and Farming*
ISBN: 07496 44389

Martin, F. *Focus on Disaster: Volcano*
ISBN: 04310 68437

Morgan, S. *Ozone Hole*
ISBN: 07496 33034

Penny, M. *Our Impact on the Planet Endangered Species (Saving our World)*
ISBN: 07502 34245

Stidworthy, J. *Earthquakes and Volcanoes*
ISBN: 1 5714 51242

Taylor, B. *Going, Going, Gone (Weird and Wonderful Guides)*
ISBN: 0-19-910836-6

Unwin, M. *Endangered Species (Saving our World)*
ISBN: 07496 3720X

Whiting, S. *Rivers*
ISBN: 07502 3377 X

Oxford Children's Encyclopedia of Plants and Animals
ISBN: 0-19-910777-7

Fiction

King *Stig of the Dump*
ISBN 0 14 036450 1

Foreman *Dinosaurs and all that Rubbish*
ISBN 0 14 055260 X

Internet

www.nasa.gov/

www.wwf-uk.org/researcher/programmethemes/rarespecies

www.mammal.org.uk/endanger.htm

Organizations

The Environment Agency
Rio House, Waterside Drive, Aztec West
Almondsbury
Bristol
BS32 4UD
Phone: 08459 333111

Friends of the Earth
26–28 Underwood Street
London
N1 7JQ
http://www.foe.co.uk

Scottish Environment Protection Agency
SEPA Corporate Office, Erskine Court
Caste Business Park, Stirling
FK9 4TR
http://www.sepa.org.uk

WWF
Panda House, Weyside Park
Godalming, Surrey
GU7 1XR
http://www.panda.org

Greenpeace UK
Canonbury Villas
London
N1 2PN
http://www.greenpeace.org.uk

RSPB
The Lodge
Sandy
Bedfordshire
SG19 2DL
http://www.rspb.org.uk

Index